SUPER-GIGANTIC ANIMALS

BLACK RABBIT BOOKS | JOANNE MATTERN

TABLE OF CONTENTS

1

The Blue Whale

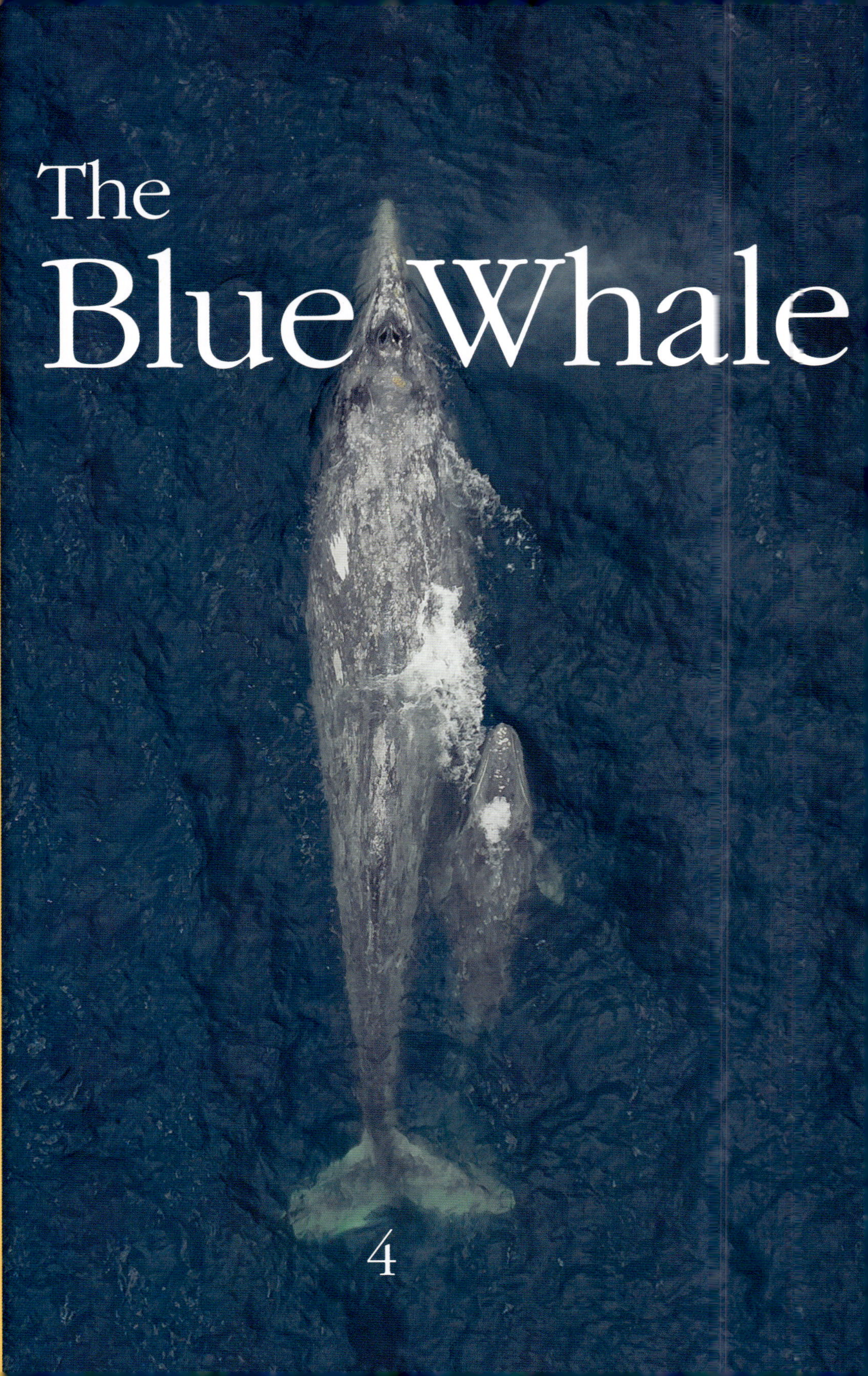

The blue whale is the largest animal that ever lived. These giant mammals are even bigger than some dinosaurs were. They live in oceans all over the world.

A blue whale is 80 to 90 feet (24–27 meters) long. It is super-heavy too! One blue whale can weigh up to 300,000 pounds (136,078 kilograms). Its heart is the size of a small car.

Sadly, blue whales are endangered. Today, there are fewer than 25,000 of these ocean giants.

Did You Know?
A blue whale's tongue can weigh as much as an elephant.

Think About It
Why are blue whales dying out? Research what factors have put these giants in danger.

2

The Polar Bear

What is the largest bear on Earth? The polar bear! It can weigh more than 1,760 pounds (798 kg). Polar bears live on the ice in the Arctic Ocean. That's near the North Pole.

Polar bears aren't just big bears. They are the largest carnivores on land. These mighty predators hunt for seals. They are great swimmers. Their feet are like big paddles. They can swim for many miles looking for food.

These big bears face big trouble. Global warming has melted a lot of sea ice. So polar bears have fewer places to live and hunt.

Did You Know?

A polar bear's skin is black. Its fur looks white because it reflects light.

3

The Ostrich

When you're 9 feet (2.7 m) tall and weigh up to 300 pounds (136 kg), it's hard to get off the ground. That is true of the ostrich. It is the world's largest bird. But it is too big to fly.

An ostrich can't fly, but it can run! These birds can race up to 45 miles (72 kilometers) per hour. They are the fastest birds on land. Their long legs can cover 16 feet (5 m) in one step. Ostrich feet make great weapons too. An ostrich's kick can kill a lion.

Think About It
If it can't fly, what does an ostrich use its wings for?

Did You Know?

The ostrich isn't the only bird that can't fly. Emus, rheas, and cassowaries can't either.

The Giraffe

4

13

The giraffe is the tallest mammal in the world. These giants are up to 18 feet (5.5 m) tall. Even baby giraffes are big. The calves are about 6 feet (1.8 m) tall.

Giraffes live in grasslands in Africa. Being tall helps them find food. A giraffe can reach leaves high in the trees. Their long tongue helps too.

Giraffes live in big herds. Staying in big groups helps giraffes stay safe from predators.

Did You Know?
Every giraffe has a different coat pattern

The Green Anaconda

5

The green anaconda wins the prize for most gigantic snake. These big creatures weigh up to 550 pounds (249 kg). That makes them the heaviest snake in the world.

Green anacondas live in hot, wet rainforests in South America. They like to lie in the water with their heads sticking up.

These big snakes eat big food. Green anacondas eat wild pigs, deer, turtles, and more. Sometimes they kill crocodiles and jaguars! These snakes are constrictors. They wrap their big bodies around their prey and squeeze it to death.

Think About It

The green anaconda is the heaviest snake. Look up the longest snake in the world. Which snake do you think is the "biggest"?

The African Elephant

6

African elephants are the largest animals that live on land. Males can be up to 13 feet (4 m) tall. Females are about 8 feet (2.4 m) tall.

These animals are super heavy too. Males weigh up to 14,000 pounds (6,350 kg). Females weigh about 7,000 pounds (3,175 kg).

Elephants are **herbivores**. That means they only eat plants. They eat roots, grass, fruit, and tree bark. It takes a lot of food to keep an elephant full. One elephant can eat up to 300 pounds (136 kg) of food a day.

Think About It Elephants are endangered too. In what ways do humans affect animal populations?

MORE TO EXPLORE

WHERE IN THE WORLD?

BLUE WHALES
All oceans except the Arctic

POLAR BEARS
Northern parts of North America, Europe, and Asia

OSTRICHES
Africa

GIRAFFES
Africa

GREEN ANACONDAS
South America

AFRICAN ELEPHANTS
Africa

MORE TO EXPLORE

FANTASTIC FACTS

A blue whale's huge lungs hold a lot of air. It can stay under water for up to an hour.

A thick layer of fat under the skin keeps polar bears warm.

An ostrich's wings help it balance when it runs.

A giraffe's neck is so long, it has to spread out its legs and bend down to drink.

Young anacondas hide in the mud to get away from predators.

One elephant can produce 330 pounds (150 kg) of poop a day.

MORE TO EXPLORE

COOL COMPARISONS

How do these animals compare in size?

Green Anaconda
550 pounds (249 kg)

Ostrich
300 pounds (136 kg)

Giraffe
3,000 pounds
(1,361 kg)

African Elephant
14,000 pounds (6,350 kg)

Polar Bear
1,760 pounds (798 kg)

Blue Whale
300,000 pounds
(136,078 kg)

MORE TO EXPLORE

RESOURCES

Glossary

carnivore (KAR-nuh-vor) A meat-eating animal.

constrictor (kuhn-STRIK-tur) A snake that coils around and squeezes prey.

endangered (en-DAYN-jurd) Close to no longer existing.

herbivore (HERB-uh-vor) A plant-eating animal.

mammal (MAH-muhl) A type of animal with hair or fur that drinks their mother's milk.

predator (PRED-uh-tuhr) An animal that eats other animals.

prey (PRAY) An animal hunted or killed for food.

Read More

Austen, Lily. *Biggest Animals*. Minneapolis: Jump!, Inc., 2025.

Bell, Samantha S. *The Biggest Animals in the World*. San Diego: Brightpoint Press, 2023.

Index

TOP RANK is published by Black Rabbit Books, P.O. Box 227, Mankato, MN, 56002. • Top Rank is an imprint of Black Rabbit Books. • Edited by Alissa Thielges • Designed by Danny Nanos • Photographs © Dreamstime/THPStock, cover; Getty Images/imageBROKER/Burkhard Sauskojus, 10, aisbeckfoto, 13, 21, Robert Harding Picture Libr. Ltd, 20; Shutterstock/cellistka, 23, chrisbrignell, 18, [illegible], Dervish45, [illegible]3, FloridaStock, 9, 21, Four Oaks, 11, gillmar, 6–7, GTW, 8, GUDKOV ANDREY, 17, Hennadii H, 23, jaroslava V, 2, 14, Marti Bug Catcher, 5, Patrick K. Campbell, 16, paula French, 12, 21, pixorena, 23, Rukqq, 23, Sam DCruz, cover, Starostov, 23, Timmy V Photography, 19, [illegible], tonyzhao[illegible]20, 15; Superstock/Amazing Aerial, 4, 21 • Printed in China

Library of Congress Cataloging-in-Publication Data: Names: Mattern, Joanne, 1963- author. | Title: Super-gigantic animals / by Joanne Mattern. | Description: Mankato, MN: Top Rank is an imprint of Black Rabbit Books, [2025] | Series: Super-incredible animals | Includes bibliographical references and index. | Audience: Ages 8–11 | Audience: Grades 2–3 | Identifiers: LCCN 2024017549 (print) | LCCN 2024017550 (ebook) | ISBN 9781645823797 (library binding) | ISBN 9781645824015 (paperback) | ISBN 9781645824237 (ebook) | Subjects: LCSH: Body size--Juvenile literature. | Animals—Juvenile literature. | Classification: LCC QL799.3 .M383 2025 (print) | LCC QL799.3 (ebook) | DDC 591.4[illegible]—dc23/eng/20240517 | LC record available at https://lccn.loc.gov/2024017549 | LC ebook record available at https://lccn.loc.gov/2024017550